Y0-AQY-314

HOMEMADE
C-4

RAGNAR BENSON

HOMEMADE C-4

A RECIPE FOR SURVIVAL

PALADIN PRESS
BOULDER, COLORADO

Also by Ragnar Benson:

Action Careers
Breath of the Dragon: Homebuilt Flamethrowers
Bull's Eye: Crossbows by Ragnar Benson
Fire, Flash, and Fury: The Greatest Explosions of History
Gunrunning for Fun and Profit
Hard-Core Poaching
Home-Built Claymore Mines: A Blueprint for Survival
Homemade Grenade Launchers: Constructing the Ultimate Hobby Weapon
Live Off the Land in the City and Country
Mantrapping
Modern Weapons Caching: A Down-to-Earth Approach
to Beating the Government Gun Grab
The Most Dangerous Game: Advanced
Mantrapping Techniques
Ragnar's Big Book of Homemade Weapons:
Building and Keeping Your Arsenal Secure
Ragnar's Ten Best Traps . . . And a Few Others
That Are Damn Good, Too
Survival Poaching
Survivalist's Medicine Chest
The Survival Retreat
Switchblade: The Ace of Blades

Homemade C-4: A Recipe for Survival
by Ragnar Benson

ISBN 0-87364-558-8
Printed in the United States of America

Published by Paladin Press, a division of
Paladin Enterprises, Inc., P.O. Box 1307,
Boulder, Colorado 80306, USA.
(303) 443-7250

Direct inquiries and/or orders to the above address.

CONTENTS

WARNING

This manual is for informational purposes only. Neither the author nor the publisher assumes any responsibility for the use or misuse of information in this book.

The procedures in this book and the resulting end product are dangerous. Whenever dealing with high explosives, special precautions should be followed in accordance with industry standards for experimentation and production of high explosives. Failure to strictly follow such industry standards may result in harm to life or limb.

"Whoever maliciously damages or destroys or attempts to damage or destroy by means of an explosive or fire any real or personal property . . .

1) Shall be imprisoned for not more than 10 years or fined not more than $10,000 or both.
2) If personal injury results, shall be imprisoned for not more than 20 years or fined not more than $20,000 or both.
3) If death results, shall be imprisoned for life or shall be subject to the death penalty."

—Federal Law Relating to Explosives

PREFACE

Survivors generally agree that commercial explosives lend themselves best to commercial applications. Paramilitary survival explosives, as a general rule, need to be more powerful. For instance, store-bought dynamite will not cut steel or shatter concrete (usually).

Many survivors believe that there are times ahead when they will need an explosive equivalent of military C-4, or plastique. However, as with the lottery, fire department, and post office, which are monopolized by various government agencies, the federal government monopolizes C-4, making it next to impossible to purchase. Survivors can't count on buying and caching military explosives against the day of need.

According to standard military charts, straight 60-percent commercial dynamite, the most powerful grade generally available to the public, has a detonation velocity of approximately 19,000 feet per second (fps). Military TNT detonates at about 22,600 fps. TNT is considered to be the minimum grade of explosive required by survivalists and paramilitarists who want to cut steel and shatter concrete.

C-4, the acknowledged big-league explosives benchmark, detonates at a speedy 26,400 fps. C-4 may seem to be ideal for your survival needs, but, as with many somewhat worthy objectives, the game may not be worth the candle. Mixing up a batch of C-4 may not be worth the risk. It is both dangerous and illegal.

Seymour Lecker, in his excellent book, *Improvised Explosives,* quotes the famous paramilitarist Che Guevara: "Fully half of the people we assigned to explosives making were eventually killed or maimed." Even the best, simplest formulas are dangerous. The one that follows is no exception. It is the safest formula that I know of, but even at that, a certain percentage of those who try to make this explosive will end up as casualties.

Federal laws regulating explosives manufacture are extremely strict. Home manufacturers can receive penalties of up to $10,000 and/or ten years' imprisonment. If personal injury to other parties results from the experiments, fines and jail sentences can be doubled.

Although there are ominous signs on the horizon, the United States does not yet seem to be part of a completely totalitarian society. In that regard, anarchy may be premature. However, this is purely a matter of personal perspective. Times and events can change quickly. Processes that may now appear unduly risky from a chemical, legal, and sociopolitical standpoint may soon be entirely acceptable. Each reader should know the risks and then apply his own standards.

If you think that you would like to have C-4 now (or possess the capability of making it at some later date), this book is for you. What follows appears to meet most survivors' specifications for a military-grade explosive. If you follow instructions carefully, the material is *relatively* safe to manufacture, but, of course, making or having it was illegal at the time this book went to press. To solve this dilemma, you may choose to master the necessary skills and store this knowledge away with the necessary ingredients in case you need them later.

INTRODUCTION

It was with intense interest that I read the account of the German police arresting a suspected terrorist in Frankfurt. The suspect, if one can believe the news accounts, was an Arab carrying a wine bottle full of nitromethane. Authorities on the scene made a great issue of the potential danger involved with possession of this hazardous material. The hapless fellow—who coincidentally was a despicable, bloodthirsty, Middle Eastern terrorist—carried the dreaded liquid in a carefully resealed liter bottle that had formerly contained red wine.

The news accounts (now three years out of date) failed to explain how the police determined that the liquid in the bottle was not, in fact, true fruit of the vine. To the best of my knowledge, customs officials—even in Third World countries—do not customarily open and sample bottles of wine taken from the suitcases of visitors crossing their borders.

But had they opened the bottle and smelled the contents, they would have discovered that nitromethane is sweet-to-neutral smelling, not altogether dissimilar to a fine port or a heavy burgundy. It definitely does not have a sharp petroleum smell as do gasoline and fuel oil. Had they tasted the liquid, the police would have been certain that it was not wine. A swallow would have killed the swallower, but not instantly. Nitromethane is not an instantaneous poison.

As a result of the article, I made a mental note of the possible future uses of nitromethane, but did nothing else. The subject came up again years later when readers wrote to me after the publication of *Ragnar's Guide to Home and Recreational Use of High Explosives* and offered several suggestions to prevent the relative listlessness I had encountered when trying to detonate ammonium nitrate soaked in kerosene.

"Try mixing ammonium nitrate with nitromethane," one especially knowledgeable former marine demolitions expert suggested. "The stuff is a real pisser," he wrote, "as fast as TNT, with just as high a brisance. It is useful for cutting steel and other paramilitary survival applications. It is about as safe a material as one can make. I would carry the separate materials in my baby's basket," he went on. "You should get to know this stuff."

Still, my extensive experience with explosives and a great deal of research had made me very conservative regarding homemade explosives. Most of them are extremely dangerous and require relatively exotic, difficult-to-obtain ingredients that should be combined under exacting conditions, preferably by a trained chemist. With the example of Silvertown, England, fresh in mind, where 57 tons of raw TNT accidentally detonated, I was understandably cautious about attempting to home-brew military-grade explosives.

In addition, although several excellent texts mentioned the use of nitromethane and the fact that it reacts vigorously with oxidizers, not one delved into using it to produce an acceptable substitute for C-4. My intellectual curiosity was somewhat piqued. There seemed to be reasonable cause to experiment with the idea to see if it could be done.

Intellectual curiosity notwithstanding, my interest in making C-4 would have remained theoretical had a young man who had recently worked for the Oregon Fish and Game Department not stopped by one day to talk about nitromethane.

"We used nitromethane to blow rocks out of streams, cut new channels, build fish ladders, and to do a bit of fish counting," he said. "We poured the premeasured nitromethane into aluminum cylinders full of finely powdered ammonium nitrate," he explained. "When we shot these charges, they were tremendously powerful, having a great shattering effect for their size and weight." The material was safe and relatively easy to handle, he concluded.

Adding everything together, I now had enough information to thoroughly attract my interest. I started working in earnest, experimenting with nitromethane, ammonium nitrate, and other easily obtainable materials. As a survivor, I felt that this information was needed by the fraternity.

The neighbors endured the smoke, noise, and flash of the hundreds of charges I prepared, but they are undoubtedly pleased that the experiments are over. In my opinion, what we have—after the smoke has cleared—is of real value to survivors. I am pleased to recommend this improvised C-4 as easy, effective, and *relatively* safe, especially if the components are stored unmixed.

CHAPTER ONE
AMMONIUM NITRATE

One may be amazed to find that something as common as agricultural-grade ammonium nitrate (NH_4NO_3) is the basis for a huge number of explosives. Ammonium nitrate is readily available on a year-round basis. Farms of every size regularly use hundreds of tons of this fertilizer.

Ammonium nitrate is often the preferred source of nitrogen for such crops as corn, wheat, beans, and barley. Farmers use it whenever they need a source of relatively stable, long-lasting agricultural nitrogen. This is especially surprising since the concentration of nitrogen per bag is relatively low, making this nitrogen source expensive for many cost-conscious farmers. Ammonium nitrate costs as much as $9 per 80-pound bag in farm supply stores and up to $15 per 60- or 80-pound bag in garden-supply stores where profit margins are steeper.

Ammonium nitrate was first produced in the early 1860s by Swedish chemists. The process they developed is the same one used today by major fertilizer manufacturers. The process entails putting natural gas under great pressure, mixing it with superheated steam, and injecting the mixture into a conversion chamber lined with a platinum catalyst. After the reaction is underway, the generated heat causes the process to be self-sustaining.

Pure liquid ammonia produced by this process is combined with nitric acid, which is also produced by most ammonium-nitrate manufacturers. (Many producers sell nitric acid to other manufacturers for use in their manufacturing operations. Although U.S. production of nitric acid and ammonium nitrate is now virtually absorbed by agribusiness, most of the plants were started with government subsidies as explosives manufacturers.) Combining nitric acid and ammonia produces salts, which after being dried and prilled should be 34 percent nitrogen.

Some fertilizers marked ammonium nitrate may actually be something else. Manufacturers often add a calcium coating to ammonium nitrate because it is deliquescent, which means it pulls moisture out of the air. Uncoated, unprilled ammonium nitrate will quickly harden into a substance resembling green concrete. Anything more than a slight calcium coating, however, will keep the activating liquid (in this case, nitromethane) from soaking into the ammonium nitrate, just as it prevents the absorption of water. If the manufacturer adds more than a minute coating of calicum, he must mark the bag appropriately. Don't use this material.

Although fertilizer-grade ammonium nitrate can usually be purchased from nurseries and garden-supply stores, a better source for explosives manufacture is farm-supply stores. Garden-supply outlets often stock fertilizers that are blends of ammonium nitrate and other fertilizers. Blends are absolutely unacceptable even if they claim to contain a base of ammonium nitrate. Buy only pure ammonium nitrate because any other additives dramatically reduce its explosive effectiveness.

Sales clerks often will try to get you to substitute urea or ammonium sulfate for ammonium nitrate. They point out that the substitute is less expensive, more stable, has just as much nitrogen, and is a prettier color. (I customarily explain that I need pure ammonium nitrate because I intend to blow up the material. This approach works best in rural stores. Urban clerks, used to supplying yuppie rose growers, may look askance at this sort of honesty.)

Would-be home-explosives manufacturers must learn to read fertilizer bags, at least in a superficial sense. The figures listed on the bag refer to the ratio of nitrogen, phosphorous, and potash contained in the product. Ammonium sulfate will be listed as 21-0-0 or something close. Urea, which can contain from 46 to 48 percent nitrogen, would read 46-0-0. Blends such as 21-44-8 contain 21 percent nitrogen, 44 percent phosphate, and 8 percent potash. These and other similar substitutes are worthless for anything other than fertilizing. Only ammonium nitrate contains a ratio of 34-0-0.

On arriving home with the 34-0-0 fertilizer (if you're not planning on using it right away), seal the unopened bag (ammonium nitrate is properly sold in plastic-lined bags, not from bulk bins) in at least two heavy-duty plastic garbage bags. Of course, any partially full bags should also be thoroughly sealed to prevent moisture absorption. Under many circumstances in the United States, it is virtually impossible to store ammonium nitrate for any length of time and still maintain usable ingredients.

Ammonium nitrate has been involved in some spectacular explosions during this century. Well over three million pounds of ammonium nitrate accidentally detonated in the harbor at Texas City, Texas, in 1947. Oppau, Germany,

Ammonium nitrate, a common garden and farm fertilizer, is available year-round throughout the United States at farm- and garden-supply stores.

was blasted right off the map in 1921 by a free-roaring ammonium nitrate blast. (For more information about these and other great explosions of history, read *Fire, Flash, and Fury* by Ragnar Benson, Paladin Press.) However, in spite of these notable accidents, ammonium nitrate is relatively safe to handle.

Many farmers store it in barns just a few feet from the house. An unlikely combination of heat and contamination by oils or coal dust can cause problems, but as a general rule, I would not be fearful of keeping the material under my bed. It is inert, as road builders, quarry operators, farmers, contractors, and others who use it as an inexpensive blasting agent find out. Ammonium nitrate must be

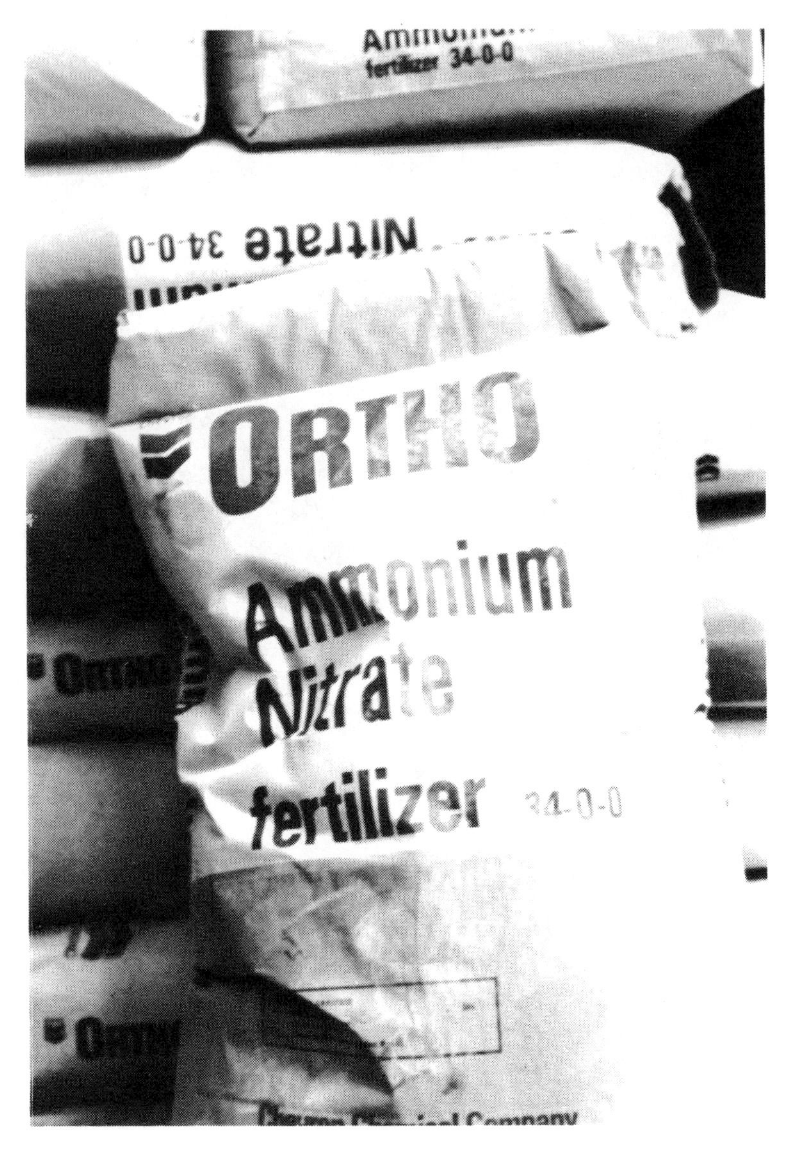

Buy only sealed bags of ammonium nitrate that is labeled 34-0-0. The figures represent a ratio of nitrogen, phosphorous, and potash. You want a fertilizer that contains 34 percent nitrogen, no phosphorous, and no potash. Don't be persuaded by salespeople to substitute other blends or products. Buy the freshest fertilizer you can find and seal it immediately in double garbage bags as soon as you get home.

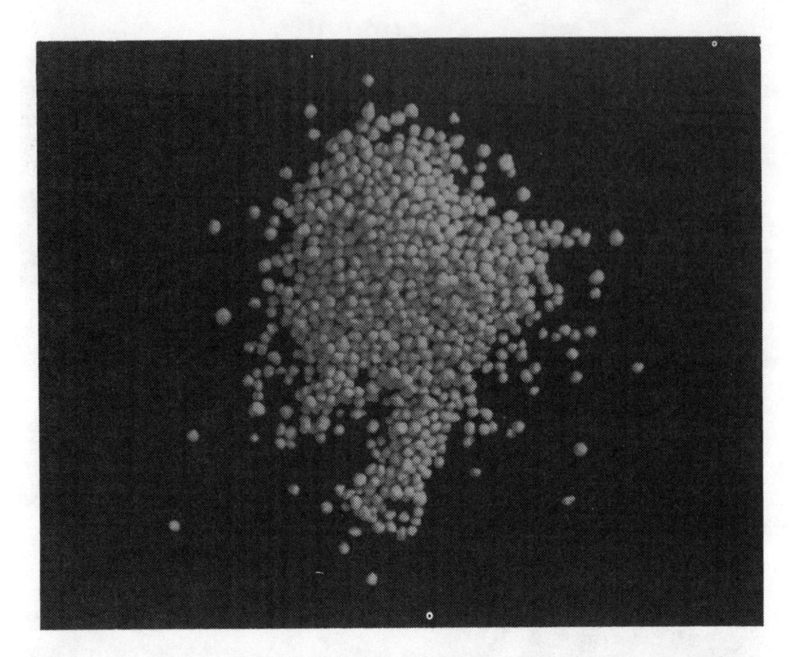

Fertilizer-grade ammonium nitrate is exuded into small seed-sized prill and then coated with a thin layer of calcium. The calcium coating is a mixed blessing. It is necessary to keep the prill from absorbing moisture and hardening into a concrete-like substance, but it also prevents the activating liquid (in this case, nitromethane) from soaking into the prill.

soaked with fuel oil and/or mixed with powdered walnut hulls, coal dust, or other source of carbon to make it active. Even with these combustible additives, I find it terribly difficult to make ammonium nitrate detonate.

Officially, ammonium nitrate is considered only a blasting agent, but it does have some explosive applications. During World War I, the British, who were low on military explosives, used a million pounds of ammonium nitrate laced with TNT and powdered aluminum to stage a successful sapper attack against the German lines at Messines Ridge in Belgium. Later on, continuing through World War II, the French and Germans both loaded their high-explosive artillery and mortar rounds with ammonium nitrate explosives.

Although many countries around the world now prohibit the sale or possession of ammonium nitrate, it is commonly available in the United States and will probably continue to be for the foreseeable future. At this time, buying an 80-pound bag should be no problem for anyone (even city dwellers) with ten dollars and a means of carting it off.

CHAPTER TWO
NITROMETHANE

Nitromethane is the second of three chemical components needed to put C-4 together in one's home chemistry lab. The material is somewhat obscure, expensive, and at times desperately time-consuming to obtain. On the other hand, it is reasonably safe to handle and can be located if one applies oneself to the task.

Nitromethane (CH_3NO_2) is used in many organic chemistry laboratories as a washing solvent and is found in virtually every college chemistry lab. Industrial firms use it to dissolve plastics, clean up waxes and fats, and manufacture numerous chemical-based products.

More commonly, nitromethane is used as a fuel additive. Model-plane enthusiasts mix it with castor oil and alcohol to power their miniature engines. It is also used to fuel small indoor race cars and go-carts. But the largest group of consumers commonly available to survivors are drag racers. It is not uncommon for quarter-milers to burn gallons of this expensive fuel on every run.

As a result, the best place to look for nitromethane is at drag strips and stock-car races. Often a local petroleum dealer will bring a 55-gallon barrel of the fuel to the track and sell it by the gallon to the drivers and mechanics. As a result, those who can't afford 55 gallons can buy enough to compete that night.

In some larger cities, petroleum dealers handle the fuel on a limited basis. An hour or two on the phone may uncover a dealer who will sell it by the gallon. Most bulk petroleum dealers will special-order a full barrel, but at $1,925 per barrel (based on $35 per gallon), few survivors would be interested.

Another likely place to look for nitromethane is in hobby shops. Most carry premixed model engine fuel, containing up to 40 percent nitromethane. Theoretically, this fuel mixture should activate ammonium nitrate, but my experience using it is mixed at best. Perhaps if the fuel is fresh and dry, it might work consistently. Yet, in spite of extensive testing, I have not achieved even a 30-percent success rate using high-concentration model fuel. The problem appears to be the alcohol which, when mixed with the fuel, pulls moisture out of the air even when the bottles are well sealed.

A few well-stocked hobby shops carry six- or eight-ounce bottles of nitromethane. Most will special-order it by the gallon at considerably more than $35 per gallon. Model-plane enthusiasts usually do not use fuel containing more than 15 percent nitromethane because it will burn up their expensive little engines. So survivors probably won't find more than a gallon or two of the high-concentration, 40 percent fuel even in well-stocked hobby shops. If they do find it, it probably will not work consistently.

If all else fails, nitromethane can be ordered at extremely high prices from chemical supply houses. Most will sell it to individuals since nitromethane does have a number of valid "civilian" uses. Check survival magazines for addresses or borrow a Fisher or Sigma catalog from the local high-school chemistry department. It may be possible to locate local industrial or commercial users who are willing to sell a few spare gallons.

Officially, nitromethane is categorized as a Class 3 conflagrant, meaning it reacts to open flame on about the same level as gasoline. It is not highly sensitive to shock. At drag strips, dealers drop barrels of nitromethane off their trucks or roll them around with impunity. They seem little concerned with the consequences of rough handling.

However, nitromethane is moderately toxic if ingested or inhaled. People who have ingested the material may suffer from nausea, vomiting, and/or diarrhea. Heavy or regular ingestion can result in permanent damage to the kidneys. Nitromethane is about as toxic and explosive as leaded gasoline in its original state.

Nitromethane is much less costly today than when it was developed at the turn of this century. Initially, it was made by reacting methyl iodide with silver nitrite. The resulting product was combined through the Kolbe reaction method, using chloracetic acid. At the time, nitromethane explosives were considered effective but far too expensive to merit large-scale production.

Today, nitromethane is manufactured by injecting nitric acid into a high-pressure chamber containing superheated methane gas, a relatively inexpensive process. At temperatures of 400 degrees Celsius the reaction becomes self-sustaining. Because its price has decreased so dramatically, nitromethane is encountered more frequently today as a fuel additive and in laboratories.

Pure nitromethane is a thin, syrupy, yellow liquid. It smells a bit sweet, but the odor is subtle enough that it is not readily recognized. Food coloring can be safely added to camouflage the liquid, if you desire. When lit, nitromethane burns brightly with considerable heat and force until the fuel is consumed. In its pure, unmixed form, it has a shelf life of about four years before moisture destroys it.

As with ammonium nitrate, possession of nitromethane is not controlled except perhaps in isolated local instances. Nitromethane can be stored by survivors for relatively long periods in plastic or steel containers. If one does not spill large amounts of the substance in an unventilated space or suck one's thumb after using it, nitromethane is relatively benign.

The challenge for survivors entails finding a source of affordable nitromethane, which may mean putting a long-term, well-programmed procurement plan into place.

CHAPTER THREE
HOME MANUFACTURE OF C-4

Making homemade C-4 requires one more chemical: denatured ethyl alcohol. This ingredient is so common and so safe that no further discussion is required—except to emphasize the importance of using *fresh* alcohol, preferably purchased from a paint-supply store.

Having come this far, most readers will agree that we are dealing with some fairly benign chemicals. Now the trick is to combine them in an effective and reasonably safe manner. As with most things in life, there is a downside. The process is not nearly as simple as one would hope, but it is possible, even for chemists with only high school training, to carry it out.

My strong suggestion remains that anyone contemplating home manufacture of C-4 thoroughly think through both the process and the consequences before he proceeds. The following procedure yields an extremely powerful explosive. It dwarfs anything available on the commercial market. Even 80-percent Hy-drive dynamite pales into firecracker class compared to the explosive you may produce.

Those who decide to proceed are also reminded that 1)

they are probably violating federal law, and 2) they should already know how to handle conventional commercial explosives competently before attempting this procedure. Experimenters should start with small test batches, remembering that those who fail to use caution, common sense, and care could face disastrous results.

Compared to manufacturing some other explosives, producing this C-4 substitute is not particularly difficult or dangerous. What danger does exist comes when combining the materials, which can be done at the last moment immediately preceding actual use.

Nevertheless, the procedures are exacting. Those who are untrained in chemistry or who are sloppy or careless will not succeed. Now that my warning is complete, let's begin.

The first step is to dry the ammonium nitrate and keep it dry. Where the humidity is high, this is a difficult to virtually impossible task.

Start by taking a one-pound coffee can or its equivalent from a freshly opened bag of ammonium nitrate. The coffee can will hold one-and-one-half to two pounds of prilled ammonium nitrate. A one-pound can provides a greater height relative to diameter, which makes the volume less dense and aids in its drying. Seal the unused bag of ammonium nitrate away in double plastic garbage bags immediately after removing the amount needed.

Place the can in an electric oven set at the lowest possible setting and dry in the oven for a minimum of three hours. Be careful that the temperature never goes above 150 degrees Fahrenheit. (Doing this properly will require a good-quality, lab-grade, dial-read thermometer available from chemical supply firms or catalogs.)

Ammonium nitrate liquefies at about 170 degrees F and will blow at about 400 degrees F. Before it explodes, it will

Scoop out two pounds of the ammonium nitrate prill into a one-pound coffee can. The height relative to diameter of the one-pound can makes the volume less dense and aids in its drying.

Dry the prill in an oven set at a low temperature (not to exceed 150 degrees F) for at least three hours, but don't let the prill melt. Ammonium nitrate vapors are toxic, so it is essential that the temperature stays low and the room is well-ventilated. On completion of the heating cycle, cap the can immediately and seal in double garbage bags. Even double sealed, the dried ammonium nitrate will absorb moisture and can be stored for no more than 12 days.

Measure exactly 250 milliliters of dried ammonium nitrate prill. The specific gravity of ammonium nitrate is 1.725, yielding a sample of 430 grams.

bubble and smoke, providing adequate warning to remove it from the heat.

On completion of the heating cycle, seal the dried prill in the coffee can and place it in double, sealed plastic bags. At most, this material will last 10 to 12 days before absorbing too much moisture—even though it is triple-sealed. Always make sure the seals are completely zipped and airtight.

Place about 250 milliliters (about 430 grams) of this oven-dried material in an ovenproof glass dish. Cover the prill with the type of denatured ethyl alcohol used to carry moisture out of gas lines (available from paint and automotive supply houses at about seven dollars per gallon).

Stir this mixture around for about three minutes or until the alcohol turns a muddy, cloudy brown. Drain off the

Wash the sample in fresh denatured alcohol to remove the remaining moisture. Buy the alcohol from a paint or auto-supply store.

Place the ammonium nitrate in a glass dish, cover with alcohol, and stir thoroughly for about three minutes.

The alcohol will remove a brown sludge from the ammonium nitrate. As soon as the alcohol turns brown, the process is completed. Throw the alcohol away.

Strain the alcohol from the prill and heat gently (I prefer an electric wok, but you can use a hot plate or stove top). Stir constantly and use an accurate thermometer to make sure temperature stays below 150 degrees F.

Heat for three or four minutes until the alcohol is completely evaporated.

alcohol by straining through a seine or screen. Dump the 430-gram sample back into the dish and gently heat over low heat. (I use a stainless steel wok at the lowest heat setting, but you could also use your stove top or a hot plate.) Use a thermometer to be certain the sample stays below 150 degrees F.

Immediately after the alcohol wash, grind the prill to avoid moisture absorption. Various methods can be used to do this. Some survivors use two flat hardwood boards, a mortar and pestle, or even an electric coffee grinder. By whatever means, reduce the prill to talcum-powder consistency.

(If the prill is not ground finely enough, it may be necessary to sieve the powder. It is hoped this step will be unnecessary. Makers will note that the ammonium nitrate begins to cake and lump from moisture when removed from the grinder. Sieving only exacerbates this situation.)

Quickly tamp or pack the powder into a container. This must be done before the ammonium nitrate begins to reattract moisture so it isn't always possible to do a thorough job. Preventing moisture absorption is your primary concern so work quickly.

When selecting a container, make certain that it is airtight. Old medicine or spice bottles work nicely. Some commercial makers use custom-made, thin-walled aluminum cylinders that look much like containers for high-priced cigars.

Although the finished product is doughlike and can be put in a plastic bag to mold around a girder or squash into a crack, it seems to have considerably more power when packed tightly in a rigid cylinder. I did not have a chronograph or any other means of measuring speed of detonation so it is impossible to make the above claim with

To grind the prill, I use an electric coffee grinder, but a mortar and pestle or two boards also work.

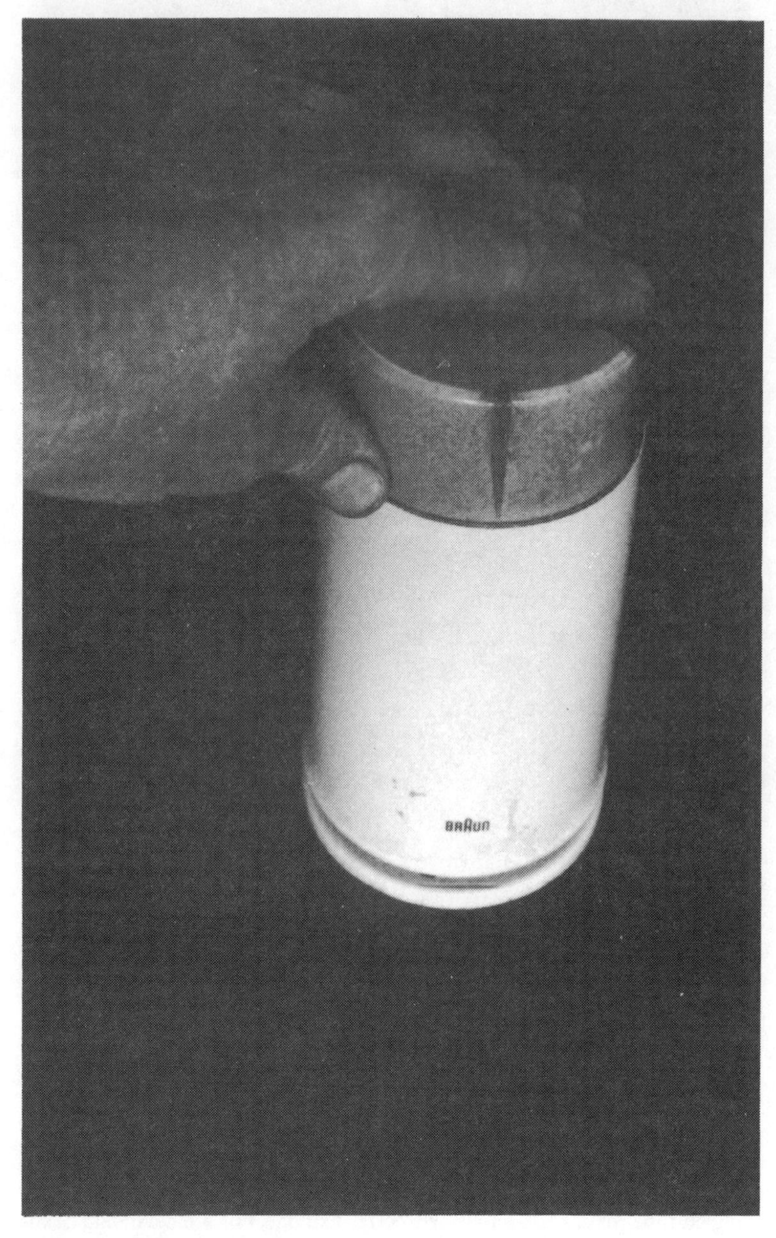

Grind the prill to a fine talcum-powder consistency.

Dump the finely ground ammonium nitrate into a solid container immediately after grinding. It is extremely deliquescent (moisture-absorbent), so seal it as soon as possible. Note the lumps starting to form immediately after exposure to air.

If the grinding process is not thorough, prill must be strained to remove coarse material. This step should be avoided if possible because it exposes the ammonium nitrate to moisture for a longer period.

certainty. However, the packed material produced larger holes in the ground because it apparently cakes better with the nitromethane when held tightly in a rigid configuration.

Whatever container is used, the maker must know exactly how many grams of ammonium nitrate it will hold. Also, there appears to be a minimum amount of powder that can be detonated. With less than 300 grams (about 10 ounces), it is tough to bury the cap thoroughly and secure a good detonation.

When deciding on container size and the amount of ammonium nitrate to use, remember to leave a small space at the top of the container for the liquid nitromethane. Using the correct amount of nitromethane to sensitize the ammonium nitrate is much more critical than one would first suppose. I avoided the need for scales by using metric measurements wherein weight and volume using specific gravity become identical.

Despite almost driving our family into poverty by my many costly experiments, I still do not feel I have all of the answers pertaining to this process. My experiments indicate that one should use slightly less than one-third nitromethane by volume, but this seems to vary from one gallon of nitromethane to the next and from one bag of ammonium nitrate to the next. Too much nitromethane will kill the mixture, while too little will not sufficiently sensitize the ammonium nitrate.

When dumped on the powdered prill, the proper amount of nitromethane will cause the powder to bubble slightly. It is almost as if there were live clams in the container, blowing in the sand after the surf rolls over them. After about two minutes of soaking, the nitromethane—*if the correct amount is added*—will saturate the powder and turn it into a thick, porridgey mass. Too much nitro will produce

Pack the dried, ground ammonium nitrate in an airtight plastic container, such as a pill bottle that will hold about 430 grams of powder and the nitromethane (430-gram charges are sufficient for most jobs survivors demand of an explosive). Solidly packed charges in rigid containers seem to have more force than charges held in loosely packed containers. Again, remember that the container for the ammonium nitrate must be absolutely airtight.

a gruel that is too thin to fire.

I used plastic pill bottles that contained about 430 grams (about 11 ounces) of powdered ammonium nitrate, and they produced very powerful blasts. A hit from this much explosive is awesome and probably sufficient to demolish small bridges, trucks, and maybe even to knock tread off a tank. Certainly in groups of two or three fired together, it would do the job.

To this 430-gram container, I added about 75 to 80 milliliters of pure nitromethane. Getting just the right amount will require experimentation. Unfortunately, I know of no formula that states precisely how much nitromethane to use. As a rough starting point, try one part nitromethane to three parts of ammonium nitrate by volume or two parts

When combined, the powder blows a few bubbles and then cakes into a tough plastic substance. The explosive will be more powerful if the caking process is undisturbed. Combine the two materials at the blast site as a precautionary measure.

Mix 80 milliliters of nitromethane into the 430 grams of ammonium nitrate. The ratio should be approximately one-third nitromethane by volume or two parts nitromethane to five parts ammonium nitrate by weight. Precise formulas must be determined by trial and error because reactions vary from sample to sample of nitromethane and ammonium nitrate.

Wait about 20 minutes for the nitromethane to soak into the ammonium nitrate. At this point, the material is cap-sensitive but does not readily detonate when dropped or shot with a firearm.

nitromethane to five parts ammonium nitrate by weight. Theoretically, the material should sensitize in five minutes, but I get better results by waiting twenty minutes.

Once the nitromethane is poured into the ammonium nitrate, there is no need to be overly concerned about moisture getting into the powder. Water would, of course, wash the mash away if it were exposed, but the plastic bottle should solve that problem. This explosive would not be the first choice for those undertaking underwater demolitions work, but it could be used if no other explosive material were available. When mixed, the shelf life seems to be a couple of weeks or more.

At this writing I am not aware of any reason—other than psychological—why this material could not be combined and sensitized ahead of time. Storing the mixed explosive

Adding powdered aluminum to the ammonium nitrate and nitromethane mixture produced this nine-inch hole in the foreground. A similar charge without the aluminum cut the seven-inch hole highlighted in the upper right corner. A comparable charge of dynamite merely skins the soft meadow ground without leaving a depression.

does not seem any riskier than storing commercial dynamite. This mixture may deteriorate in time, but my experiments did not indicate this.

Although the combined material seems safe to handle, it is definitely exciting when detonated with a number six or eight cap. Commercial dynamite detonated on bare, hard ground will skin it up a bit. This explosive will dig six- or seven-inch holes without top tamping of any kind.

I estimate the velocity of detonation to be about 21,000 fps or slightly less than TNT, which detonates at about 22,600 fps. C-4, the explosive benchmark, roars out at an incredible 26,600 feet per second. The additional speed between commercial dynamite at 19,000 fps and C-4 is what cuts steel and shatters concrete. One is for homeowners, the other for survivors.

Recounting, to make C-4:

1. Use fresh NH_4NO_3.
2. Dry the NH_4NO_3 in an oven at low heat (less than 150 degrees F) for three hours or more.
3. Wash the NH_4NO_3 in alcohol until the alcohol turns muddy brown.
4. Dump the prill in a metal container and dry them thoroughly over low heat (less than 150 degrees F).
5. Grind the NH_4NO_3 as fine as talcum powder.
6. Pack a premeasured amount in a rigid airtight container.
7. Pour in one-third nitromethane by volume.
8. Wait twenty minutes.
9. Shoot with a cap similar to dynamite.

It is important that all of the steps be undertaken carefully and methodically, and that one experiments before going out in the field with military objectives in mind.

CHAPTER FOUR
THE FINISHED PRODUCT

We stood back about 90 yards from the small 11-ounce dab of explosive as the fuze slowly burnt its way down to the cap. In our experience, 90 yards was more than sufficient to protect us from such a small amount of explosive.

My many failed experiments with this material had left me uncertain as to whether we had anything more than another dud. The mountain meadow behind my cabin was strewn with ruptured plastic containers, left by dynamite caps that failed to detonate the explosives.

This time when the detonation hit, it was spectacular. A successful blast at last! The last time I experienced anything similar, I was firing LAW rockets at Fort Benning, Georgia. I vividly remember when the concussion from the three-pound warhead thumped us, even at 200 meters. I also remember a similar reaction while running through the army's live-fire tank-commander school south of Boise, Idaho.

Although I lacked sophisticated test equipment to measure its impact, the explosion undoubtedly had sufficient

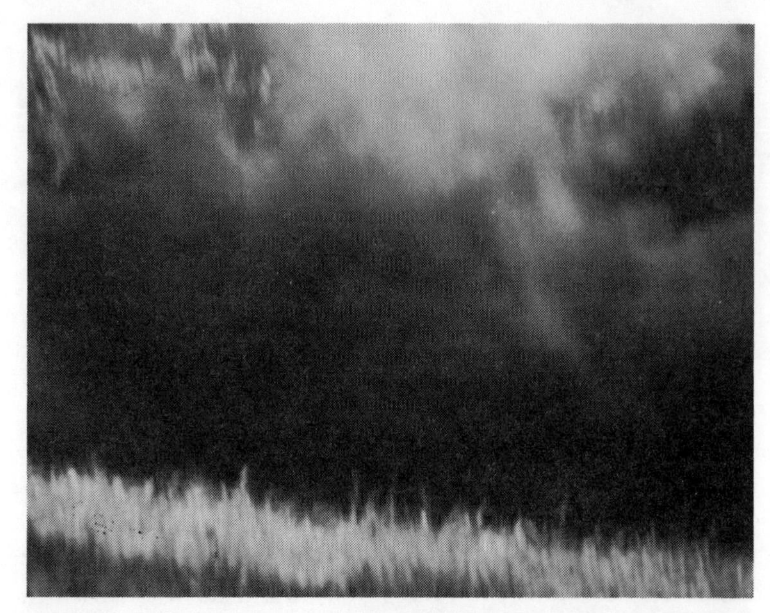

Detonation of 11 ounces of the homemade C-4 produces an impressive explosion. Experienced powder monkeys who witnessed the explosion agreed that this material is much faster than commercially available explosives.

brisance to cut steel and shatter reinforced concrete. Several observers with military experience agreed that the homemade C-4 was formidable.

The afterglow from my original success kept me going when my next few attempts turned out to be duds because my ammonium nitrate had become water-soaked. I blew my materials budget, but eventually the results became consistent. The process produces the following reaction: $NH_4NO_3 + CH_3NO_2 = H_2O + CO_2 + NO_2$!

As a practical explosive, this material seems ideal. Two shots fired from a high-power rifle do not tell the entire story, but smacking the explosive with my .223 at 45 yards did not produce a detonation. To further test its sensitivity, I set a batch aside for a week. Then I threw it down a rocky ledge and later burned it on a bed of logs without any apparent effect. Even the burning itself was not particularly notable.

This explosive is remarkably similar to genuine C-4—particularly in its stability—but it lacks one of C-4's more desirable attributes. The brisance of this improvised C-4 was not as great as that of the genuine article. It wasn't off much, but the last 5,000 fps might mean the difference between a good and an excellent explosive. Boosting this material into the C-4 class became my goal once the secret of consistent manufacture was in hand.

The tip-off to a possible solution came while I was researching World War I's Messines Ridge sapper attack. Messines Ridge was the only actual trench-warfare offensive sapper action during a war that was fought almost entirely as a set-piece contest. After 18 months of preparation, the nine tunnels filled with almost one million pounds of explosives were detonated on June 7, 1917. The resulting blast was heard by British Prime Minister David Lloyd George from his home in London 200 miles away.

Britain's World War I explosives manufacturers added finely ground aluminum powder to this explosive, called ammonal, to boost its brisance. Ammonal was used because two years of protracted warfare had consumed virtually all of Britain's conventional explosives. It was manufactured using 72 percent ammonium nitrate, 12 percent TNT, and 16 percent finely ground aluminum powder.

Having made that discovery, I began to experiment with powdered aluminum. I added it to the ground ammonium nitrate before adding the nitromethane. At a level of about 5 percent (or about 20 grams) mixed thoroughly into 430 grams of NH_4NO_3, the effect was dramatic. Instead of seven-inch holes in the earth, I was gouging out nine-inch

The detonation speed of the homemade C-4 is about 21,000 feet per second (fps), much faster than commercial dynamite but slower than TNT. The addition of finely powdered aluminum will boost the detonation speed nearer to that of genuine C-4. You can buy powdered aluminum at paint or chemical supply stores. Powdered aluminum radiator sealers work well.

Quickly and thoroughly mix the powdered aluminum into the aluminum nitrate prill *before* adding the liquid nitromethane.

craters with less than three-fourths of a pound of explosive!

Fine-ground aluminum powder is available from well-stocked paint stores and chemical supply houses, but the best place to buy it is from an automotive-parts shop. It is used to plug leaky radiators and is sold in 21-gram tubes.

Some aluminum powder is too coarse to enter into the detonation reaction. But most samples are finely ground and, for the price, work quite well (about $13.85 per pound). Purists can obtain very finely ground aluminum flakes from chemical supply houses if use of this relatively expensive (from $30 to $40 per pound) material seems warranted.

Theoretically, it would be advantageous to pack the explosive in small plastic bags that could be molded around a piece of steel or other object that one wished to cut. What scant printed information is available on this explosive suggests that the material should remain undisturbed and unmixed after the addition of the nitromethane.

Without careful, controlled testing, we do not know if the combined materials become dangerously sensitive after mixing. So as a precaution, take to the blast site carefully premeasured amounts of aluminum powder in small sealed tubes and similar containers of premeasured nitromethane to pour into the powder. Inserting the cap and placing the charge should take about 20 minutes, and the charge should then be ready to do its work.

Although this process is not unduly threatening to those who have handled explosives, it is an exacting and mostly untested one. Those who do not carefully follow all instructions should expect dangerous or poor results. Those who proceed with intelligence, caution, and diligence can expect to produce an explosive that will make despots tremble in their boots.

CONCLUSION

Other materials exist that can be combined with ammonium nitrate to produce high-grade explosives. Some quite powerful ones aren't as deliquescent as nitromethane, giving the impression that they might be more desirable than nitromethane. One formula that is currently making the rounds among survivors involves mixing two parts of NH_4NO_3 with one part hydrazine. The resulting liquids are reportedly the most powerful chemical explosive known to man—short of an actual atomic reaction.

An almost insurmountable problem with this explosive is the fact that anhydrous hydrazine is extremely corrosive and therefore desperately difficult to handle. It will blister an animal's lungs with just one diluted whiff. Professional industrial chemists use moon suits, respirators, and supplemental air supplies and still are very reluctant to do any more than a minimum amount of work with this chemical. Eventually it will eat through virtually anything metallic, making it almost impossible for survivors to store it at home. Unvented hydrazine fumes kill very cruelly in a matter of seconds.

As a result, the material is almost impossible to ship. Most carriers justifiably do not want to handle it, and partly as a result, it is also extremely expensive to purchase. It usually costs about $100 per pound, but that does not include shipping. Furthermore, it cannot be sent by United Parcel Service, Federal Express, or parcel post. So home chemists must drive hundreds of miles to pick it up personally or pay trucking charges of up to $25 or more per shipment.

It is quite possible that three pounds of finished explosive using hydrazine could cost $150 or more. When combined, the resulting liquid is extremely corrosive, toxic, and shock sensitive. I know of no storage container that would hold the material. It can't be metallic and, if a glass jar ever broke or spilled, cleanup might assume catastrophic proportions.

As a result, it doesn't take a Phi Beta Kappa in chemistry to conclude that the ammonium nitrate/nitromethane mixture is superior for survivors' purposes—despite a slightly diminished brisance. In addition, hydrazine products require the use of sophisticated laboratory equipment not usually available to survivors. Buying this equipment could make the overall cost of the project prohibitively expensive for most budgets.

For the process recommended in these pages, one needs only common household items: a set of ovenproof glass dishes; a standard measuring cup; a standard probe thermometer; a coffee grinder; an electric wok; and a tea sieve. There is no need for extra-large glass beakers to handle the reacting chemicals, lab-accurate stainless thermometers, ice baths, air-evacuation equipment, or moon suits and respirators.

After nitromethane and ammonium nitrate are combined,

the mixture is reasonably safe and can be handled by most people, whereas hydrazine is too unstable to carry around or combine at the job site. Fumes from the reaction could poison everything downwind for several hundred meters. It also might arouse people's suspicions to see survivors running around in moon suits and respirators.

Other formulas for making C-4 substitutes abound, such as mixing pure nitric acid with glycerin to yield nitroglycerine. Nitric acid is obtainable and can be handled by amateur chemists, but it is somewhat risky.

Homemade nitroglycerine must be washed and purified to an extent that taxes the skills of sometimes chemists. Impure nitroglycerine grows increasingly sensitive on the shelf until simply moving the container could cause premature detonation. After my reading through detailed

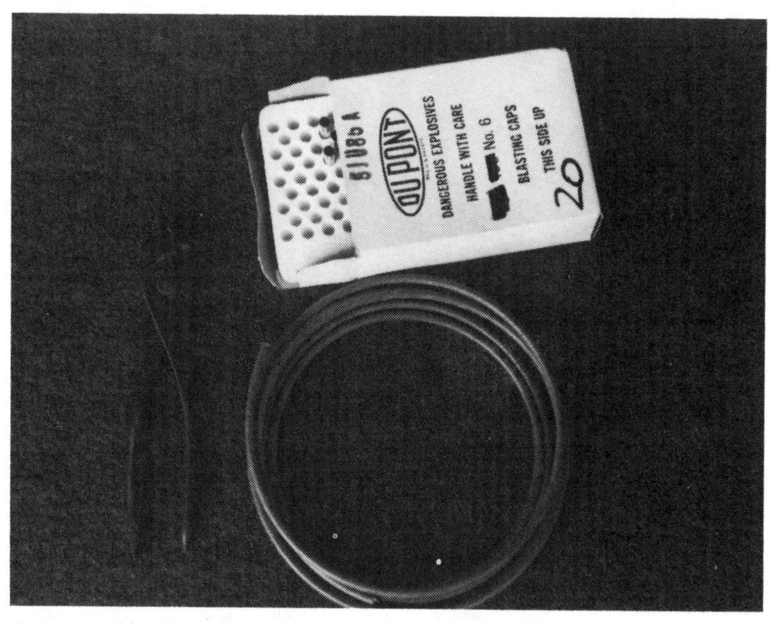

People unfamiliar with commercial explosives should not attempt to make C-4 at home. Use standard commercial caps to detonate the charge.

manufacturing instructions, it was easy to conclude that this process is unnecessarily difficult and dangerous.

In summary, the explosive made by mixing ammonium nitrate with nitromethane seems to possess all of the desirable characteristics of high-grade military explosives that are otherwise unavailable to survivors. The process has few disadvantages that I have been able to identify.

Note:

Readers will note that throughout this discussion I have assumed the use of commercial safety fuze and caps or standard electrically fired dynamite caps. This book assumes that makers already know enough about explosives to know where to purchase the necessary caps and fuze.